PETROGLYPHS OF HAWAI'I

by
Likeke R. McBride

Petroglyphs of Hawai'i
by Likeke R. McBride

To
IVA McBRIDE
my mother

ISBN 0-912180-60-9

Cover Illustration by
EDWIN KAYTON
www.kayton-art.com

Full Color Cover designed and printed by
DON O'REILLY, Hilo Bay Printing
Back cover illustration created using "Petroglyph Pictofont" published by Guava Graphics, Honolulu, Hawaii

Map Illustrations by
KATHRYN GODSHALL

published by the
PETROGLYPH PRESS, LTD.
160 Kamehameha Avenue • Hilo, Hawai'i 96720
Phone 808-935-6006 • Fax 808-935-1553
BBinfo@BasicallyBooks.com • www.BasicallyBooks.com

CONTENTS

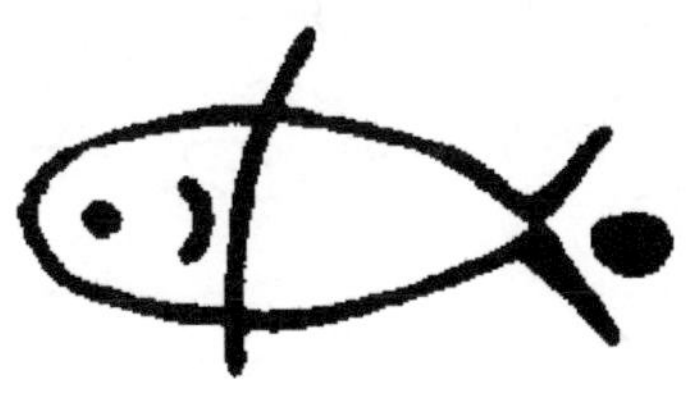

McBride

FORWARD

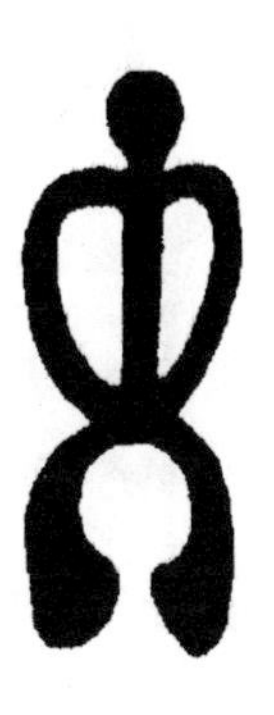

When *Petroglyphs of Hawai'i* was first published in 1969 there was very little in print on the subject. In the ensuing years, research has added a great deal to our knowledge of Hawaiian life in prehistoric and historic times. In addition, researchers are much more sensitive to the sacred places and traditions of the Hawaiian people.

The author, Likeke R. "Dick" McBride passed away in 1993 and regrettably could not be consulted on this revision. The author's son, Andrew S. McBride, provided valuable assistance in the revision, calling upon his experiences accompanying his father in the field as McBride conducted his research, as well as offering his skills in editing and proofreading.

The publisher has chosen to reproduce the original text with only a few notations, additions and corrections. This was deemed necessary in order to provide readers with more up-to-date information concerning the location, treatment and study of Hawaiian petroglyphs. For instance, McBride used the term *kaha ki'i,* translated as drawn or scratched picture, for petroglyphs. However, the most recent edition of the Hawaiian-English dictionary lists the term *ki'i pohaku,* stone image, as the more appropriate term. For this reason the text has been revised to reflect the more common terminology.

Since 1969, the development of the island of Hawai'i and the incursion of lava into petroglyph

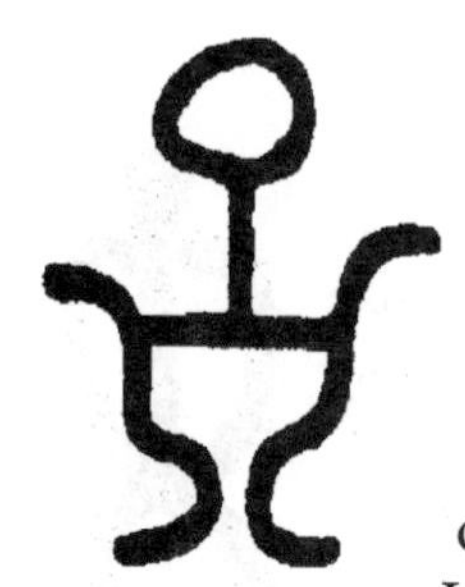

areas have made changes to some of the more important petroglyph sites in Hawai'i. The continued eruption of Kīlauea volcano has covered up the petroglyph site of Pu'u mana wale'a (hill of rejoicing) and hundreds of petroglyphs along the coast within Hawai'i Volcanoes National Park and continues to threaten the famous Pu'u Loa petroglyph field. Resort development in the South Kohala and North Kona areas has greatly impacted the formerly remote petroglyph sites at Puakō, 'Anaeho'omalu and Ka'upulehu. To the credit of certain developers, some care was taken to preserve the cultural heritage of the surroundings as the resorts were planned. However, vandalism to petroglyphs in these areas is on the rise, pointing to the need for education concerning the treatment of petroglyphs.

It was formerly common to make rubbings or even castings of petroglyphs. Because this causes permanent damage to the images these practices can no longer be encouraged. For this reason, it was decided to delete the information concerning rubbings and castings that was found in the original text. Most of the sites are located in extremely dry areas, where even using chalk to define a petroglyph may leave an impact for years to come. Natural erosion coupled with the impact of visitors simply walking through have taken a toll on many of these sites.

New information concerning Hawaiian petroglyphs continues to accumulate and public interest in petroglyphs has never been stronger. Researchers have long taken an interest in the subject. One of the best sources of information is *Hawaiian Petroglyphs,* which was first published in 1970. This informative book was written by J. Halley Cox and Edward Stasack, who were both with the art department at the University of Hawai'i. Cox and Stasack combined ethnographic accounts with their own observations to propose some probable functions for petroglyphs in Hawai'i. Stasack has continued to play an active role in the study of petroglyphs, recently recording sites on the islands of Hawai'i and Kaho'olawe.

Since 1980, Georgia Lee, an art historian at UCLA, has been active in recording and documenting petroglyph sites in Hawai'i and Easter Island. Her 1999 book, *Spirit of Place,* is the most definitive work to date on the study of Hawaiian petroglyphs. Lee's emphasis is on both documenting the petroglyphs and determining why they were only created in certain locations. This includes her observation that petroglyphs exhibit regional styles that may provide clues into specific functions for petroglyphs produced in different locations. For instance, the profusion of sails at Ka'upulehu indicates to her that sailing canoes were of special import there, possibly representing a school for sailing or navigation. Lee has also developed a typology of Hawaiian petroglyphs for a computerized data base of petroglyph sites and individual motifs.

The study of petroglyphs is also of interest to archaeologists

in Hawai'i. The specific patterning of petroglyphs along boundaries, trails and sacred sites, provides clues to archaeologists into their possible function as markers and use in sacred ceremonies. Stylistic patterning of motifs may also eventually aid in developing regional petroglyph typologies for individual islands. In addition, new dating techniques have proven invaluable in estimating when petroglyphs were made. This aids archaeologists in the relative dating of associated archaeological sites (i.e. such as a house site that is built over a petroglyph field).

Stasack and Cox developed a relative dating technique that is unique to petroglyphs in Hawai'i. The age sequence is based on the superimposition of petroglyphs and places linear motifs as the oldest, followed by the triangular figures and muscled forms. Recently radiocarbon dating of petroglyphs has been successfully accomplished by using organic matter that has accumulated sequentially inside the petroglyph since it was first made. This method (AMS 14C varnish dating) first attempted in Australia, has been successfully used to date petroglyphs on the islands of Kaho'olawe and Hawai'i and promises to yield more dates for petroglyphs in the islands. The "best estimate" for the oldest petroglyphs dates back to AD 983.

Petroglyphs are an important resource to all people of Hawai'i and care should be taken when walking near them. Tread lightly and carefully when visiting petroglyph sites, and treat them with respect and aloha.

By Christine Reed, Publisher and
Catherine Glidden, M.A., Archaeologist

INTRODUCTION

The word petroglyph is derived from the Greek roots, *petros* - stone and *glyphe* - carving. The term especially applies to pictures and symbols cut into a rock surface by the people of prehistoric time.

Some of the oldest of these found are the engravings of glacial age animals carved in European caves 10,000 years or more ago. The soft rock of the cavern walls permitted full exercise of the artists' ability to realistically reproduce bison, mammoths, deer and horses hunted for food during that time. Some of these etchings were perhaps merely for the decoration of sacred places or living sites. Most, however, appear to have been for the purpose of insuring good luck to the hunters, upon whom depended the livelihood of the entire group.

The world's petroglyphs belong largely to the Stone Age, but since the knowledge of metals arrived in different places at different times, the Stone Age may be said to be nearly coincident with the length of time man has existed.

In 1627 Peder Alfson, a professor, wrote a letter to Ole Worm, the father of prehistoric studies, describing the rock drawings in the province of Bohuslan in southwestern Sweden. Included with his letter were rubbings of the rocks that constitute the earliest reproductions of Bronze Age petroglyphs. The figures carved in the rock depict vessels, animals, carts, weapons, and the humans who inhabited the southwestern coast of Sweden between 1000 B.C. and 500 B.C. Recently it has been demonstrated that they are a part of a culture spread entirely across northern Asia to the Pacific.

Petroglyphs of Hawai'i is not intended as a definitive study of a forgotten art. Almost anything written about petroglyphs seems assured of raising more questions than answers. Perhaps, like much of our art today, they could only be explained by the artists who made them.

Swedish Petroglyphs
Hawaiian Petroglyphs

HAWAIIAN PETROGLYPHS
KI'I POHAKU

Pu'uloa was a village
Papalauahi a sleeping place
A shed for Kīlauea
When Pele came in the night
Tossing and turning the humpback waves.
ancient chant

Although petroglyphs can be found on all the Hawaiian islands, by far the most numerous and accessible sites, like Pu'uloa, are located on the island of Hawai'i.

Pu'uloa (long hill) is not a village today and can barely be termed a hill, but it is "peopled" by a multitude of two-dimensional figures called petroglyphs carved into its smooth *pahoehoe* lava surface. Today, when Pele, the Hawaiian goddess of volcanoes, sends rivers of molten rock down the *pali* from the flank of Kīlauea, archaeologists, artists, Hawaiiana buffs and art lovers all hope together that the petroglyph field of Pu'uloa will be spared.

Petroglyphs are scattered throughout more than 100 places in the Hawaiian Islands and constitute the only prehistoric art not owned by private collectors or enclosed in museums. Because they remain out in the open where they were made, they are particularly vulnerable to vandalism, the elements and, on the island of Hawai'i, being covered by lava flows.

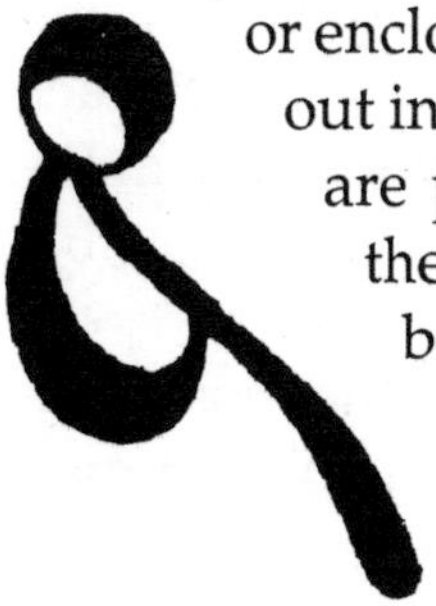

In the Hawaiian language the petroglyphs are called *ki'i pohaku*. *Ki'i*, meaning picture or image and *pohaku,* stone. Literally meaning "image in stone." Petroglyphs were made in different ways depending on the hardness of the rock surface to be worked upon. Dr. Kenneth Emory of the Bishop Museum recognized three methods, which he called pecking, bruising and abrading. Pecking or hammering on *pahoehoe* lava with a dense, hard beach pebble breaks up the cells or vesicles of the surface and can produce a deep design.

On a glazed surface, such as the lining of a lava tube, bruising or gentle scraping effects a color change. Abrading is hammering and cutting the surface of the lava with a filing or scraping motion.

The term *kaha ki'i* may be applied to pictures scratched in the sand. In Hilo, Hawai'i, between Coconut Island and Leleiwi Point, is a beach named Onekahakaha, which means "picture drawing sand." It has undoubtedly changed a great deal in appearance since the people of old scratched pictures there.

Drawing was probably so commonplace in old Hawai'i that it is rarely mentioned in tradition. It is used as a point of explanation in one of the oldest stories concerning the islands. In the legend of Aukelenuiaiku, it becomes necessary for him to leave his homeland and find a new place to live. On the beach his grandmother draws a map in the sand showing him the far places beyond the sea.

Sand was probably the most common drawing medium in olden times. While a piece of coral or a sea urchin spine scratched on black lava gives great contrast, it is difficult to correct and almost impossible to erase. A drawing in sand was temporary, of course, but the sand could be used over and over. When permanence was desired, stone was probably the material used. Tradition tells us that the *heiau* (temple) was first designed in sand before a stone was moved toward its construction.

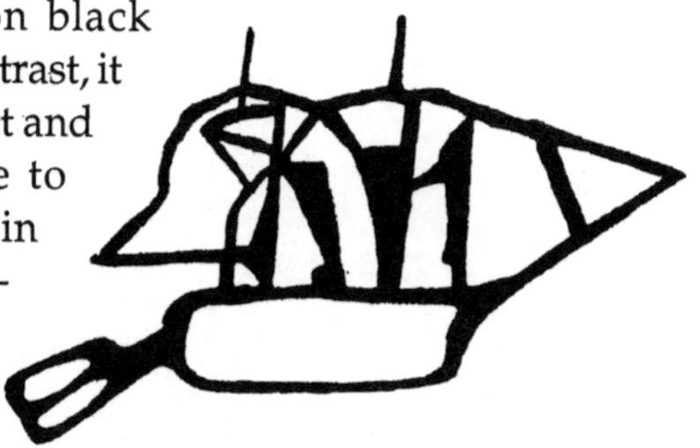

Drawing pictures in the sand was still a popular amusement for small boys during the early 1800's. John Papa Ii, in his *Fragments of Hawaiian History*, tells of a day in his youth spent with two companions making sand drawings of the sailing ships in Honolulu harbor.

While sand sketches were probably made *pi 'ani wale no* (just for play), there were other times perhaps when pictures were desired for a more serious purpose, such as permanent decoration. In the collected articles of S. M. Kamakau titled *"Ka Poe Kahiko,"* the author discourses on the disposal of corpses in ancient times. "There is only one famous hiding cave, *ana huna*, on O'ahu. It is Pohukaina... This is a burial cave for chiefs, and much wealth was hidden away there with the chiefs of old... Within this cave are pools of water, streams, creeks, and decorations by the hand of man *(hana kinohinohi'ia)*, and in some places there is level land."

Petroglyphs? Perhaps.

A CANNIBAL KING AND PETROGLYPHS

The dismembering of Captain Cook in 1779 by the natives of Kealakekua, Hawai'i, had a profound effect on many of the early visitors to the Sandwich Islands. Rumors were rampant that the inhabitants had cooked the captain and eaten all but his hands, which had been returned to the British.

Cannibalism was not unknown to the Hawaiians, but their acquaintance with it was through stories handed down from an earlier age. As remarkable as it seems, the search for a legendary cannibal's gravy dish led to the discovery of the first petroglyphs reported in Hawai'i.

In 1822 an inquiring traveler named G. F. Mathison was told the story of Aikanaka (man eater), a chief who, with some of his people, came to the islands in olden times from a distant land. These foreigners landed on the island of Kaua'i and lived there until the Hawaiians discovered their strange appetites and they were driven away. Aikanaka and his party then took up residence on the island of O'ahu at Waialua and called this place

Halemano (shark house). The cannibal king and his followers committed such atrocities there that the natives exterminated them in a terrible battle.

At the end of the story, Mr. Mathison was told that the place called Halemano still existed, but that all that remained was the *imu* (underground oven) and the cannibals' *ipukai* (gravy dish).

The visitor was taken to see these fragments of the past and wrote in his *Narrative of a Visit to Brazil, Chile, Peru and the Sandwich Islands,* "I had expected to find a monument of great magnitude; instead of which I saw nothing but a flat stone, resembling an English tombstone, about five feet broad by six or seven in length. The surface was very smooth and upon it I discovered many rude representations of men and animals, similar to those, which have from time to time been met with and described among the Indians of America. Many were defaced and in others I could trace no resemblance to any known objects, either animate or inanimate; the stone itself was very imperfect, pieces of it having evidently been broken off on different sides, which I learnt from the guide had been done by the neighboring inhabitants in order to convert the materials into knives, mirrors, pots, and other domestic utensils, which were always fabricated from stones in former times, previous to the introduction of iron by foreign traders. Annexed is a drawing taken on the spot."

Mr. Mathison states, "The habitation of the said Chief was situated on the very spot since called after him Herimino, where I now stand, and the stone in question served as an altar upon which the unfortunate human victims were sacrificed. Near it is a large round hole, about twenty feet in circumference, and still clearly discernible, which was pointed out as the place where the *kanakas*, or men, were cooked and devoured by the Chief and his adherents."

The year after Mr. Mathison visited O'ahu, the missionary Reverend William Ellis made his famous hike around the island of Hawai'i. In the appendix to the journal of that trip, published in 1826, he included a few lines concerning rock carvings. "...Along the southern coast, both on the east and west sides, we frequently saw a number of straight lines, semicircles, or concentric rings, with some rude imitations of the human figure, cut or carved in the compact lava..."

Public interest in Hawai'i's petroglyphs began with an article titled *"The Pictured Ledge of Kaua'i" by* J. K. Farley that appeared in the 1889 issue of *Thrum's Hawaiian Annual.* Mr. Farley speculated that the drawings must be very old since the beach had apparently subsided at least six feet and the petroglyphs were nearly always covered with sand. Twice in a decade he had seen them uncovered, revealing seventy-six pictures and

After sketches by Farley and Judd

markings up to six feet in length.

Mr. Farley's article seemingly sparked the search for rock pictures. Within a few years A.F. Judd and John Stokes found petroglyphs on Moloka'i, O'ahu and Kaua'i. Dr. Kenneth Emory found them on O'ahu, Maui, and Lana'i. C.M. Walton and W.D. Westervelt reported finding drawings in caves on Hawai'i.

In 1906, John Stokes located the acres of petroglyphs at 'Anaeho'omalu on Hawai'i's South Kohala coast and, twelve years later, Reverend Albert S. Baker found the extensive petroglyph field near Puakō only a few miles farther north.

D.K. Reed

Petroglyphs found at Puakō.

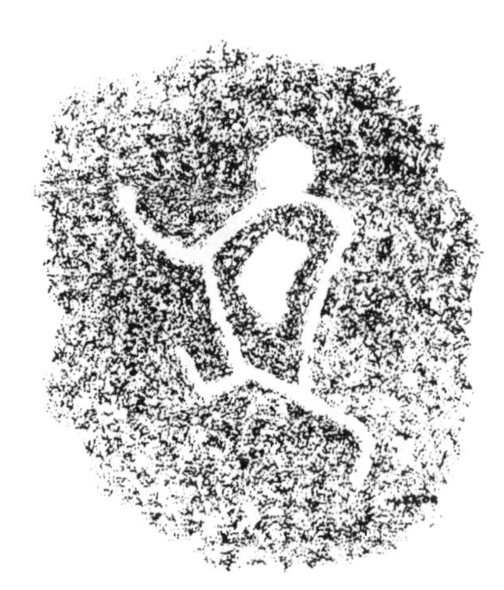

WHO MADE THE PETROGLYPHS?

It seems reasonable that nearly all of Hawai'i's petroglyphs found so far were made by Hawaiians. For the most part, they appear to depict people, tools, animals, vessels and weapons known to the natives before and after the arrival of Captain Cook. Granted that the petroglyphs were made over a long period of time, where did the idea for making them come from?

One of the persistent concepts of the origin of the petroglyphs is that they comprise a part of the Polynesian heritage brought along in the great migrations into the Pacific from their ancestral homeland in southern Asia. If this surmise was correct, we might be able to trace the movements of the ancient Polynesians by the petroglyph trail they left behind. Unfortunately, perhaps, such a track does not seem to exist. A rough line drawn from Hawai'i to New Zealand apparently divides the Pacific into two petroglyph provinces. On the islands west of the line petroglyphs are rare or absent, while on the islands east of the line petroglyphs are common. This suggests that if Hawai'i's petroglyphs are not indigenous, perhaps their origin lies to the southeast of the Hawaiian Islands. It is in that direction, on Easter Island, that both petroglyphs and a written language existed in prehistoric times.

According to Hawaiian tradition, the idea of images was brought to the islands by foreigners. Pele, the very fair woman from a distant land that was apotheosized as volcano goddess, is credited by some legends as bringing the first. When she and her party landed on the southern coast of the island of Hawai'i, she fabricated an image for the people of that place. It was to that same shore that Pa'ao, a foreign priest, brought two gods, one large and one small, probably three-dimensional figures.

The introduction of petroglyphs by seafarers has long been an attractive premise to scholars and enthusiasts. Around the turn of the century, a popular belief was that Indians from North America had created the rock drawings. Mr. Farley noted that Kaua'i's Pictured Ledge was a common depository for logs of redwood and fir floated from the western United States. He suggested that native canoes could sail before the wind or easily drift from the continent to the islands in a short time.

Thomas G. Thrum saw a resemblance to Native American sign writing in the petroglyphs found on the floor of an O'ahu cave. He related a story of long ago in which a skin canoe manned by brown foreigners commanded by a chiefess landed near Waimea, O'ahu, and the people remained on the island. In his 1915 *Hawaiian Annual*, Mr. Thrum drew a comparison between local petroglyphs and those of Glen Canyon, Arizona, pictured in the August 1914 issue of *National Geographic* magazine. "The similarity of the work and form design is no mere coincidence. The resemblance is so strong that but for the distance of separation from the continent one would say at a glance that the crude drawings were all made by one and the same artist."

There is, of course, the possibility that petroglyphs might have been introduced into North America and Hawai'i by another agency. Both history and archeology indicate that the Pacific Ocean has been crossed since ancient times.

Recently it has been established that the Jomon culture of Kyushu, Japan, was transported to Ecuador, South America, about 3000 B.C., perhaps by accidental drifting. Could that landing be the foundation of the Peruvian legend that ". . . in the beginning white men came. They were tall and bearded, wore white robes and taught all manner of peaceful things"?

In 449 A.D., Captain Hwi Shan reported to his emperor in China that beyond the sunrise sea he found a land he had named Fu Sang and had returned with rocks and plants of that place and had started the true religion there.

When a contingent of Coronado's expedition marched up the eastern shore of the Gulf of California in 1540 A.D., they found a port where two vessels lay at anchor. The ships were small, square-rigged, and had golden pelicans on the prow. Around pavilions set up ashore were a few kinky-haired men and others with long straight hair, brunettes, who indicated that their country lay beyond the ocean sea toward Asia. Near the mouth of the Colorado River, the Spaniards found foreigners were working mines and extracting metals.

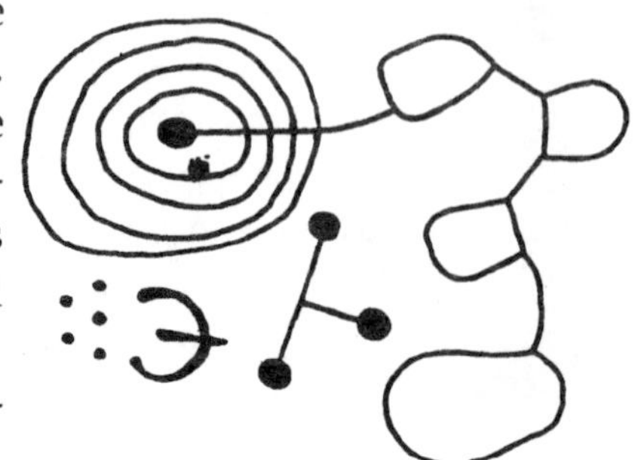

The unwritten litera-

ture of the Hawaiians is replete with tales of foreigners coming to the islands. That the Oriental and Spanish vessels may have sailed in the latitude of Hawai'i makes many of the legends more credible.

Some of the earliest strangers to arrive were the *Ehu*. They were said to be fair-skinned people with red hair, who brought with them the concept of printing tapa cloth instead of painting it.

During the lifetime of Opili, the son of Pa'ao, a group of white men landed on the southwestern side of Hawai'i. As high priest, Opili was asked by the king to deal with the newcomers. He had all manner of foods cooked and placed in nets for carrying. Then he caused great bamboos to be cut and a white square of tapa cloth fixed to the end of each. The food and banners were carried to where the strangers hid and after they had eaten, Opili advanced and spoke in the foreigners' own tongue.

There is a legend of a fair-skinned man and his sister who were the only survivors of a shipwreck. On the *kona* side of Hawai'i they knelt a long time on the beach and the place ever after was called Kulou, bowed head.

When Ahoukapu was king, seven foreigners dressed in white or yellow landed at Kealakekua, Hawai'i. Their painted boat had neither mast nor sails but had a canopy over the stern. One of the men wore a sword and had a feather in his hat. They married island girls and finally became ruling chiefs on Hawai'i. Among the treasures shown to Captain Cook in 1779 was what appeared to be the point of a broad sword.

Until modern times seamen have been notoriously

unlettered and few ships carried more than one or two literate persons. Nevertheless, there are generally a few of the crew that can draw familiar things, such as ships, animals, flags or an emblem that might identify their homeland. It is surprising that of all the strangers that may have visited the Hawaiian Islands in the past, none left a recognizable sign. However, even if the coat-of-arms of a European family was found carved in an old lava surface it would not eliminate the possibility that the local petroglyphs developed independently in Hawai'i.

Artifact in
Trocadero Museum, Paris
from "Arts and Crafts of Hawai'i"
by Te Rangi Hiroa (Peter H. Buck)

Petroglyph at
Puakō, Hawai'i
Courtesy of Paul Rockwood

WHAT DO THEY MEAN?

According to some Hawaiians, the petroglyphs were made by priests and had meaning to the initiated. Many of the designs are identical to tattoo motifs depicted by early visitors, including figures, animals, birds, concentric circles and sometimes unfinished forms.

In 1824, a high born Hawaiian woman named Kapi'olani, with a vast following of relatives and retainers, trudged more than a hundred miles from the Kona side of Hawai'i to the summit of Kīlauea where the volcanic fires flamed unceasingly. A priestess of Pele confronted her there with a piece of *kapa* from which she began to recite the edicts of the volcano goddess. When Kapi'olani in turn looked at the paper the priestess held and couldn't comprehend the symbols, she denounced it as a fraud, and went on to defy Pele and declare herself a Christian.

The implication that the priesthood of old Hawai'i had the means of a kind of writing is interesting. Few people, perhaps, would have been more surprised than William Ellis, who, only a year before, had probed the natives in regard to their knowledge of that ability. In his journal, he put down, "[Writing] also appears to them a most surprising art...Supposing it beyond the powers of man to invent the plan of communicating words by marks on paper, they have sometimes asked us, if, in the first instance, the knowledge of it were not communicated by God himself."

In Ellis' opinion, the petroglyphs were "the first efforts of an uncivilized people toward the construction of a language of symbols." He was, perhaps, one of the first Europeans to ask the meanings of the rock drawings he encountered on his trip around Hawai'i. Concerning these he recorded, "On inquiry, we found that they had been made by former travelers, from a motive similar to that which induces a person to carve his initials on a stone or tree, or a traveller to record his name in an album, to inform his successors that he has been there."

This implies that some of the people had a mark or a recognizable design which, like a signature, could be identified by other people. We are told by S.M. Kamakau that the great chief of Hawai'i, Lonoikamakahiki, had as his emblem a *ka'upu* bird. One wonders if it was a marked device as well as a bird skin fastened to the mast of his vessel as a flag of privilege. Dr. William Ellis, the assistant surgeon with Cook's third expedition, commented on the tattoos the people wore on Maui. "Both sexes have a particular mark according to the district in which they live, or is it rather the mark of the *ali'i* or principal man under whose immediate jurisdiction they are." On the same voyage, at Ni'ihau, Captain Cook remarked that "one of the men had the figure of a lizard punctured on his breast and upon those of others were the figures of men badly imitated." An old Hawaiian questioned by Mr. Kramer in 1899 had a row of seven birds tattooed from breast bone to

shoulder. The native said these were *koa'e,* the white-tailed tropic bird. The figure is identical to a symbol found in the large field of petroglyphs at Pu'uloa today.

The majority of petroglyphs seem to be single figures cut into conspicuous places along a trail. Almost all of the human representations are male or at least few are undoubtedly female. While many things Hawaiian are pictured, there are others that are notably rare or absent, such as trees, shells, whales, and drums. Almost every weapon in use is figured, including clubs, daggers, slings and weighted tripping ropes.

Some of the individual petroglyphs seem to fit the descriptions of the king's men in olden times given by N.B. Emerson.

Puali -- soldiers tightly belted with the *malo,* which they wore rather tighter than was the custom among the common people. Hence the name *puali,* cut in two, from the smallness of the waist.

Uhaheke -- means with the thighs bent, consequently on the alert. They are contrasted with those who squat down on the ground. They generally carried some weapon concealed about them.

Lava -- men of great strength; only a slight interval between ribs & hips.

Kuala-peho -- men powerful with the naked fist. Strong hands.

The extensive petroglyph fields on Hawai'i are located on what is, or once was, a searing expanse of glaringly smooth *pahoehoe* lava. These are the dry *kaha* lands which could be interpreted to mean "picture place" and at the same time refers to legendary hot desolate shores. It may be just a coincidence, but hardly a better place could be selected where rock carvings on exposed lava would be so little damaged by time and the elements. Other locations where rock drawings are found may be reflected in the place names of the areas: Pu'u Ki'i, Kahalu'u, and Ki'i, meaning image.

Hawaiian Man with Gourd Mask
by John Webber, artist on Captain Cook's Third Voyage, 1778-1779.

It has been suggested that the petroglyphs were made by travelers who stopped to rest and whiled away the time by pounding doodles in the rock surface. Two or three of the petroglyph fields on Hawai'i were within a half hour walk of a settlement where water and shade were available. It seems unlikely that sensible people needing rest would stop on a hot, treeless lava flat and engage in a tiring hour of carving a picture for fun.

Perhaps a more acceptable idea is that the petroglyphs were a kind of sympathetic magic dealing with a belief the Hawaiians had in the power of words and in the power of place. In olden times the district of Puna on the island of Hawai'i was considered a propitious place to begin something. *Puna* means, among other things, beginning. It was a favorable place for a king to begin the year, for a festival to commence, or a concept to be advanced. Again, to wear a *hala lei* on an important venture might prevent success since the word *hala* means not only the pandanus, but also failure.

Petroglyph at Puakō, Hawai'i
Courtesy of Paul Rockwood

The extensive petroglyph field in Puna is at Pu'uloa - literally, "long hill." According to Sam Konanui, a Hawaiian who lived only a few miles away, Pu'uloa also carries a figurative meaning of longevity. For that reason it was a favorite location for a baby's navel cord to be placed, in order to give the child the best chance for a long life. It is interesting to speculate that travelers might also have cut their "signatures" there as a wish for long life or that the place might have served as a Hall of Fame where the "names" of famous athletes or soldiers of whom the people were proud would live forever.

Some of the most interesting petroglyphs are those which appear to be grouped in an orderly fashion. These are found mainly on the island of Hawai'i and are thought to picture events or perhaps tell a story. A few are enclosed by a carved groove reminiscent of the line or cartouche surrounding a king's name in Egyptian hieroglyphs.

One of the most intriguing petroglyphs of all seems to include a Hawaiian vessel, concentric circles, and a Hawaiian word or name in English letters.

The word *punohu* means to billow out, as a ship's sail, to spread out, as a shrub with low branches or as a cloud, (as ripples?). Is this coincidence or could it be a sort of Rosetta Stone by which the meanings of other petroglyph groups might be determined?

The complete understanding of Hawaiian rock pictures will probably never occur. Of course, it is unnecessary to fathom their meanings to enjoy them. The primitive art of the petroglyphs can be appreciated for its aesthetic value alone, but it is difficult not to wonder who made them, for what purpose, and when.

DATING THE PETROGLYPHS

In old legends of Hawai'i, the dry, hot shores where the large petroglyph fields are found are called the *kaha* lands. Since one of the meanings of the word *kaha* is to mark or draw, there is a possible implication of vast antiquity to at least the beginnings of the petroglyphs.

When J. K. Farley was examining the pictured ledge of Kaua'i, he found that strangely, there was no native tradition of the work or workers. An old Hawaiian woman named Kauila, who lived near Keoneloa Beach for many years, was interviewed. She said that she had first seen the pictures in 1848, at the age of thirteen, when she and her classmates had gone to the beach with their teacher, a Roman Catholic priest.

"We saw all the picture rocks exposed," she said. "You have seen only a part of them today. Another ledge from fifty to one hundred feet further inland, under the sand, has pictures of birds, fishes, a canoe and strange animals cut on it. The animals are not like anything now seen; they have bodies like cattle, heads and ears like pigs, but no horns; the canoe has no outrigger or figures on it. The priest went home with me from Keoneloa and talked with my father and grandfather, also with a number of old natives about the drawings. They had all seen the pictures, but had never heard who cut them or

why they were done. The oldest folks said that their fathers and grandfathers had told them that the pictures had always been there."

Although other puzzles of antiquity such as stone walls, roads, and fish ponds are ascribed to the *menehune*, a legendary race of small people sometimes mentioned as the original inhabitants of the islands, surprisingly, the petroglyphs rarely are!

There are, of course, many petroglyphs that can be dated with reasonable accuracy. The Hawaiian names and words were probably cut into the rock sometime after the 1830's when schools were established to teach the natives to read and write English. Somewhat older, perhaps, are those that picture ships, horses, goats and flintrock rifles. Some of the latter may date back to the time of Captain Cook or beyond if it should sometime be proven that Europeans arrived earlier.

If the petroglyphs were made over a considerable length of time by similar people, it is conceivable that they would evolve, that is, that the older would be more simple and the younger more complex. Dr. Kenneth Emory demonstrates this evolution by using petroglyphs of the human figure. The line or stick figures are the simplest and apparently the oldest. The triangular-bodied figures are more complex, hence younger, and the realistic, muscled figures the youngest. The hypothesis was tested by applying the laws of position and superimposition. In a large area of

Evolution of Petroglyphs

petroglyphs the oldest pictures should form a nucleus or center and the petroglyphs added later should be generally placed around them. On a single boulder, for example, the drawings might be crowded together and often overlapping. With patience and luck, one can sometimes determine which are the older figures. If a weathered or indistinct figure has another superimposed upon it that is distinctly cut, then reasonably, the youngest is the more discernible.

Applying these principles, Dr. Emory's hypothesis concerning the evolution of the human figure drawings appears valid. The relative ages of other less common pictures may be justifiably construed in the same manner. The dog that is drawn as a stick figure is apparently older than the dog made in outline or a completely cut out petroglyph. Unfortunately, these rules cannot be applied extensively to other figures because some are unique and others widely separated.

Curiously, the oldest type of petroglyphs seem the most common, the triangular figures less numerous, and the muscled, realistic figures seem relatively rare. Does this apparent decline indicate an increasing lack of interest in making rock carvings?

Another method of trying to date the petroglyphs is attempting to associate them with some legendary happening in the time of a Hawaiian king or chief listed in an accepted genealogy. However, most of the traditions concerning petroglyphs were forgotten long ago, if, indeed, many ever existed. In 1906 John Stokes found some rock carvings at

Kahalu'u, Hawai'i, that were said to represent a former king of Maui.

Kamalālāwalu

At Kahalu'u, Mr. Stokes talked with an eighty-six year old Hawaiian named Malanui who was the grandson of the last priest of Kapuanoni Heiau in the vicinity. The old man led him to the beach and pointed out a petroglyph he said was Kamalālāwalu, king of Maui, when Lonoikamakahiki ruled Hawai'i during the last part of the sixteenth century.

According to legend, Lonoikamakahiki and his brother, Pupuakea, once visited Kamalālāwalu. Pupuakea was general of Hawai'i's armies in spite of his very short stature. Kamalālāwalu decided that such a small general couldn't be much of an opponent and pondered plans to take over the island of Hawai'i. The Maui king's spies erroneously reported only a fraction of the big island's manpower which persuaded Kamalālāwalu to ready his fleet to wage war. When they invaded Hawai'i near Kawaihae, Lonoikamakahiki and his brother were at Kahalu'u. A plan formulated by two old priests was put into effect by Hawai'i's king. He deliberately disgraced the priests and drove them away. They then fled to the invading army of Maui and persuaded its leader that if he drew up his forces near Waimea, he could certainly crush Hawai'i's armies. The ruse was effective, and in the subsequent battle, the king of Maui was slain.

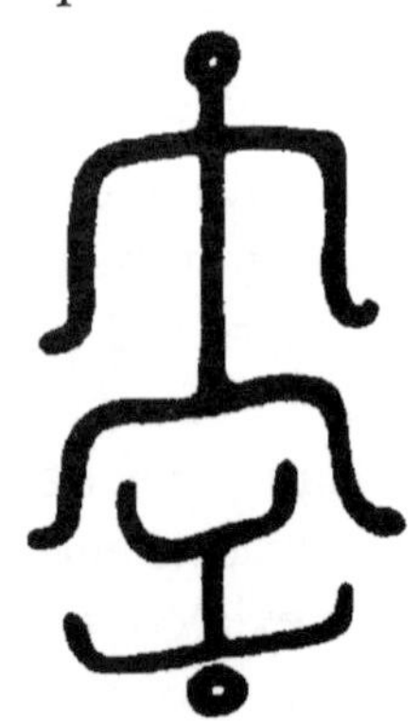

Malanui told Mr. Stokes that the

body of Kamalālāwalu was brought to Kahalu'u, a picture of it made on the rock, and the body sacrificed in the nearby *heiau* of Ke'ekū.

Perhaps four centuries before Lonoikamakahiki, there came to the islands a foreign priest named Pa'ao. He was said to come from Kahiki, the ancestral homeland. At that time a harsh leader named Kamaiole was king of Hawai'i. Pa'ao engineered a successful rebellion against the tyrant and Kamaiole was killed, which left the island without a proper ruler. The chiefs then wanted Pa'ao to be king, but he declined, saying his was the work of the gods. He did, however, send for Pili who was made sovereign and began the line of kings that extended to Kamehameha. Tradition tells that Pa'ao not only brought a new religion with elaborate temples, but also introduced the idea of images.

An interesting facet to this story is that Kamaiole was reputedly buried at 'Anaeho'omalu, a place famous today for its tremendous field of petroglyphs.

Only a few miles away, near Puakō, is one of the places where Pele, the goddess of volcanoes, purportedly landed when she first arrived in Hawai'i. That fair woman from a foreign land is also credited with bringing to the Hawaiians the concept of images. Pele and her red-haired family may have come to the island a few generations before Pa'ao.

Perhaps as interesting a method as any for guessing an early date for Hawaiian petroglyphs is the change in sea level. During the last two thousand years, the average rise in sea level throughout the world has only been about three inches every hundred years. On the western coast of Hawai'i there is good evidence that the rise of sea level relative to the land is far greater; about twelve inches a century. This increase is due to the island slowly sinking in isostatic adjustment to the great weight of the volcanoes.

Assuming the petroglyphs at Kahalu'u were carved well above the mean high tide line where sea erosion might be slight, then those under water today could be four feet or more below their original level. If that estimate is reasonable, and the apparent sea level is rising a foot a century, it would indicate that the petroglyphs there were made about 1570 A.D. or before. This date seems consistent with the legend of Lonoikamakahiki.

The older, more northern islands of the Hawaiian chain are believed to be partly compensated isostatically and the apparent sea level rise would not be so great.

In his study of the petroglyph ledge of Kaua'i, J.K. Farley mentioned that the beach had obviously subsided at least six feet. If the change in sea level alone was responsible, it could indicate a date of about 640 B.C. which is so far unsubstantiated by archeological evidence.

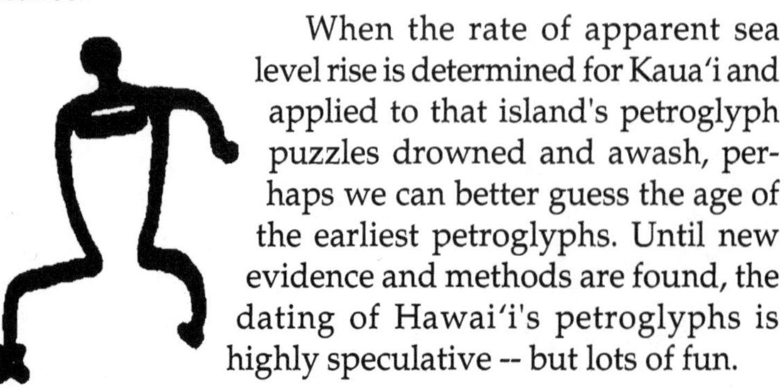

When the rate of apparent sea level rise is determined for Kaua'i and applied to that island's petroglyph puzzles drowned and awash, perhaps we can better guess the age of the earliest petroglyphs. Until new evidence and methods are found, the dating of Hawai'i's petroglyphs is highly speculative -- but lots of fun.

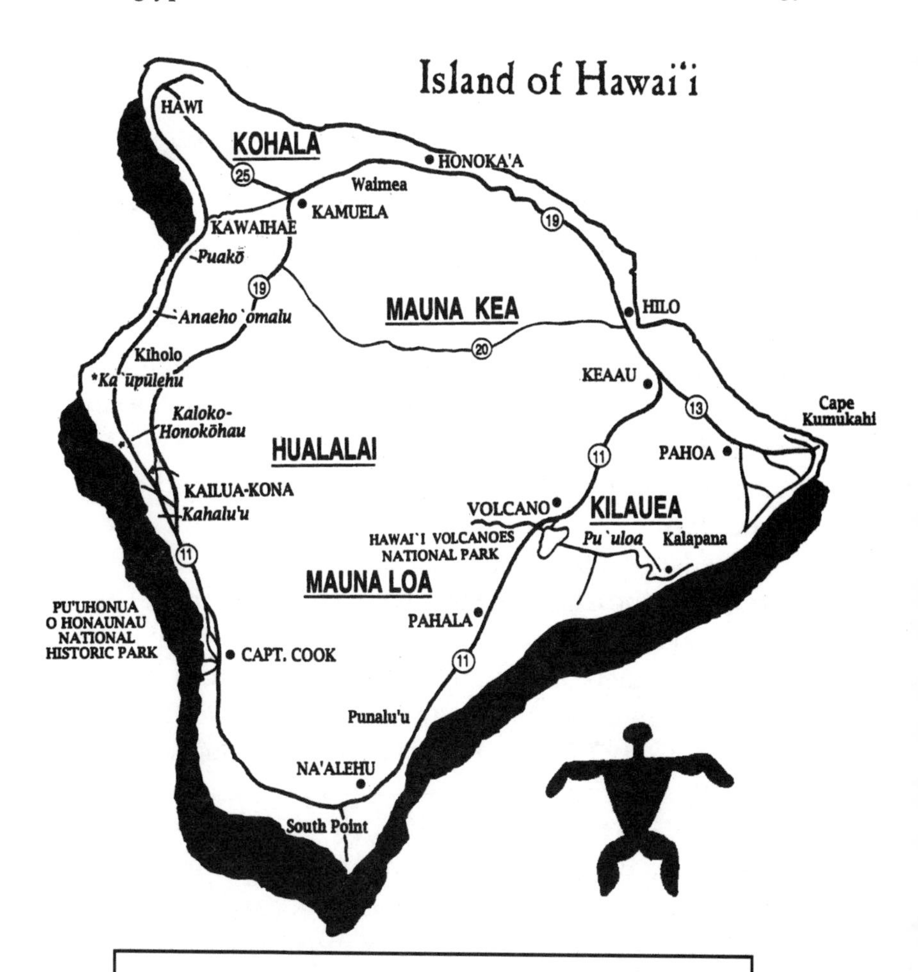

By observing a few common sense rules the life of the petroglyph may be extended.

Rock surfaces are fragile, weathered and crumble beneath your feet: avoid walking on the petroglyphs. Improperly done rubbings can discolor and damage the rock.

By observing from the boardwalk, you are helping to preserve them for future generations to enjoy.

The sign placed at the site of the Pu'uloa petroglyphs states well how petroglyphs should be viewed.

PETROGLYPH PLACES

There is little wonder that many of the early travelers to the islands do not mention seeing Hawai'i's rock pictures. The conspicuous ones seem to be in out-of-the-way places, while those that are readily accessible go unnoticed unless they are pointed out. Of the multitude of visitors that have stopped to look at the huge lava block in the lawn in front of the Hawai'i County Library in Hilo, only a few have mentioned the petroglyph deeply carved in the upper face.

The large tabular rock is the Naha Stone, reputedly brought to the island of Hawai'i from Kaua'i by the high chief Makaliinuikualawalea in ancient times. He transported the sacred stone by double canoe from its resting place near Kaua'i's Wailua River to Pōnahawai, where the city of Hilo stands today, and placed it near the temple of Pinao.

The Naha Stone was said to be used in a test to determine if a baby was legitimately of Naha rank or royalty, but it gained the fame it has today as the rock was overturned by Kamehameha the Great to prove he could rule all of Hawai'i's islands.

By far the largest concentrations of petroglyphs are to be found on the island of Hawai'i. Several sites are readily accessible and present informative interpretations while also attempting to protect the petroglyphs from damage beyond the natural weathering that oc-

From a cave near Pahala, Hawai'i

curs over time. The billowing pahoehoe lava flows offered a readily carveable surface for the artists. The hard basalt surfaces on which petroglyphs are found on the other islands required different techniques. Many petroglyph sites throughout Hawai'i are located on private property. Permission may be required to reach them. Please be respectful of both the place and the landowner's rights.

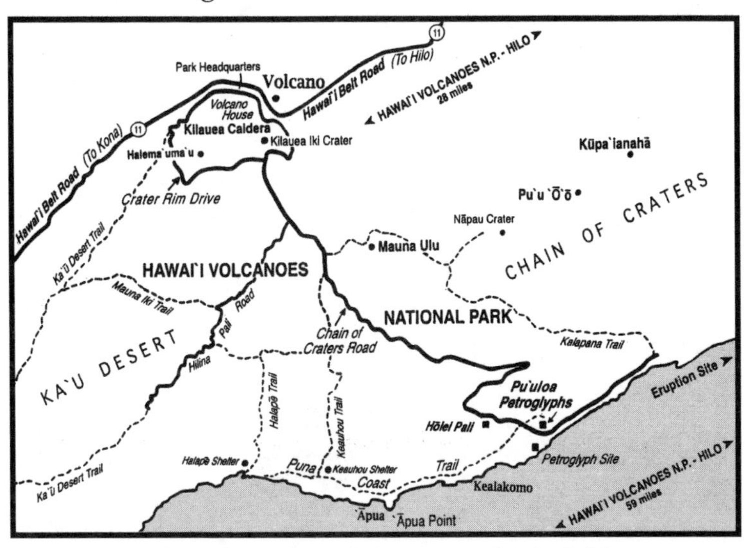

Directions to Pu'uloa petroglyph field

HAWAI'I

PU'ULOA - The Pu'uloa site boasts the highest concentration of petroglyphs in Hawai'i. This area is reached by driving through the Hawai'i Volcanoes National Park and down the Chain of Craters road toward the present eruption site. Watch for the Emergency Telephone signs about 19 miles from the park entrance. The old coastal trail used by Hawaiians is a reddish trace winding between billows of *pahoehoe* lava on which you can walk to the petroglyph area, about .7 mile (about a twenty minute walk). A wooden boardwalk has been

PETROGLYPHS AT PU'ULOA, HAWAI'I

constructed to afford visitors an elevated view of the petroglyphs while preventing further damage to the fragile images. Take along plenty of water -- it can be desiccatingly hot and dry. If you follow the 'Āpua Point trail across the road from the parking area toward the sea, you can find petroglyphs within a five to ten minute stroll.

The circles and dots appear to be the oldest petroglyphs and are thought to be the depository of the *piko* (navel cord) of a baby to take advantage of the figurative meaning of Pu'uloa (long life). Others may be a record of a trip around the island, as described by Rev. William Ellis in his journal. "When there were a number of concentric circles with a dot or mark in the center, the dot signified a man, and the number of rings denoted the number in the party who had circumambulated the island. When there was a ring and a number of marks, it denoted the same; the number of marks showing of how many the party consisted; and the ring, that they had travelled completely round the island; but when there was only a semicircle, it denoted that they had returned after reaching the place where it was made."

Many visitors may have passed this place in olden times on a pilgrimage to the volcano summit by way of the sea-cliff village nearby named Kealakomo (the way to go in). In Hawaiian tradition, when a person recovered from what was believed an incurable illness, he

On a boulder at Kamo'oali'i, Hawai'i

From a cave near 'Āinahou Ranch, Hawai'i

PETROGLYPHS AT PUAKŌ, HAWAI'I

See pages 19, 53-56 & 58 for photographs of Puakō Petroglyphs.

made a "journey of health." This trip included walking through part of Puna, climbing the steep path to Pele's house on the top of Kīlauea, and swimming around Mokuola (island of life), now called Coconut Island in Hilo Bay.

The eruption of Kīlauea volcano at Pu'u O'o which began in 1983 has sent a series of lava flows down the slopes and into the sea, covering the Chain of Craters road on the Kalapana side of the park. The destruction of the visitor center and the *heiau* at Kamoamoa was a great loss. Lava flows long spared the Waha'ula Heiau, a *luakini heiau* or temple of human sacrifice. Meaning "red or sore mouth," it was the last temple to hold public worship after the abolition of the *kapu* system. On August 12, 1997 the heiau was entirely covered by lava. As the eruption continues steps are being taken to map and catalog the Pu'u Loa petroglyph field to preserve a record of this culturally rich area, should future lava flows move in a westerly direction and cover the site.

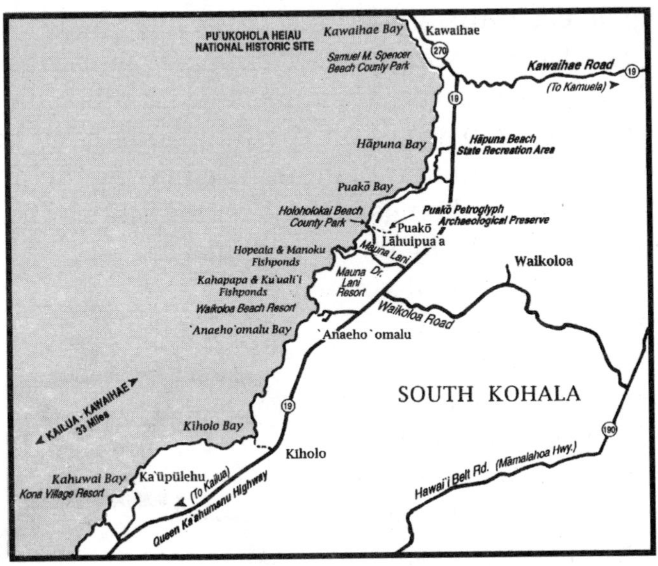

Location of petroglyphs at Puakō and 'Anaeho'omalu

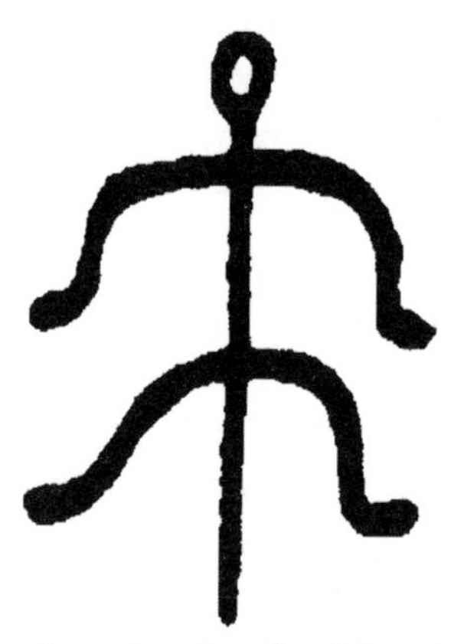

PUAKŌ -- The Puakō Petroglyph Archaeological Preserve in South Kohala is no longer accessible from the village of Puakō. The Mauna Lani Resort at Kalāhuipua'a has done an admirable job of providing an opportunity for those interested in petroglyphs to enjoy and learn more about them while protecting one of the largest petroglyph sites in the Pacific. An informative brochure and map of the area is available at the resort.

To reach the site, travel to the end of Mauna Lani Drive, 2.4 miles from Highway 19. Watch for the entrance to the Holoholokai Beach Park on the right. The park is open 6:30 am to 7 pm. The Malama trail winds through *kiawe* forest .7 miles to the petroglyphs. There you will find an elevated viewing area with a trail circling an enclosure around one of the most highly concentrated areas. Appropriate clothing, footwear, water and sun protection are recommended for this trail.

Numerous petroglyph reproductions have been placed at the beginning of the trail for making rubbings. Please confine rubbings to these reproductions and refrain from disturbing the ancient images.

The pictures that comprise the preserve are widely scattered along both sides of the ancient *Kāeo* trail. One meaning of *kāeo* is the winning, which is reminiscent of the petroglyphs made on Easter Island to commemorate the annual

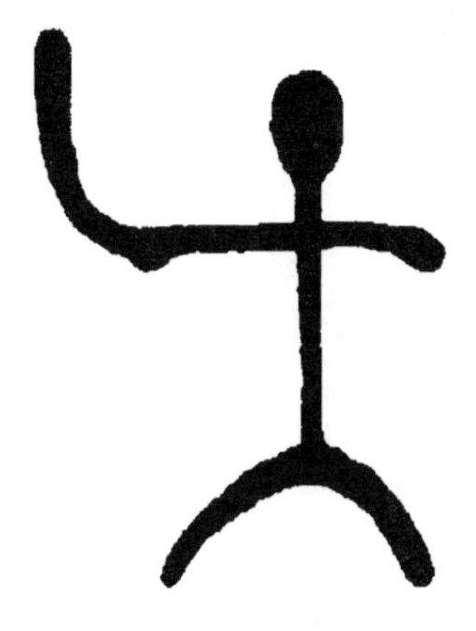

winner of the contest to find the first frigate bird egg.

In general, the petroglyphs of Puakō are thought to be some of the oldest on Hawai'i because most of the images are of the linear or stick figure type. One of the interesting groups of these human representations is a file of thirty men each above the shoulders of another (see photo page 54). Many of the other drawings appear to be *koa* (soldiers) wielding various weapons. An old Hawaiian riddle asks, "What is the tree that goes to war?" The answer was, "Koa, which means both a tree and a soldier."

During the ten years' war for Hawai'i, King Kamehameha was opposed by Keoua, a military strategist from the Ka'u district. When the war ended about 1790, Keoua was sacrificed at Pu'ukohola Heiau near Kawaihae and later buried at Paniau, Puakō.

'ANAEHO'OMALU -- This petroglyph field is about four and a half miles south of Puakō, on land developed into the Waikoloa Beach Resort area. Drive .5 miles from Highway 19 to the parking lot at the King's Shops. Information and maps for a self-guided walk are available there or from The Royal Waikoloan and other hotels in the vicinity. It is a short trek to the petroglyphs located adjacent to the golf course.

The Reverend William Ellis probably landed near this place, but, unfortunately, did not see it. Here are many enclosed drawings along with groups of others that may tell a story. The circles and dots described by William Ellis, are also common at 'Anaeho'omalu.

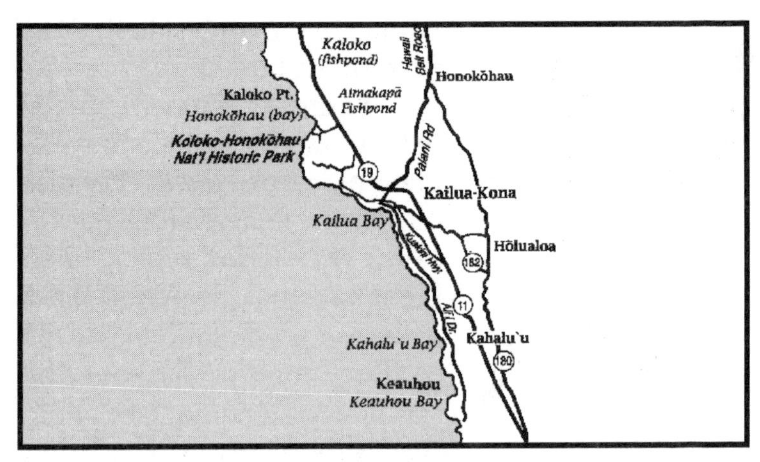

Location of petroglyphs at Honokōhau and Kahalu'u

KA'UPULEHU -- Many unusual petroglyphs are found here on the grounds of the Kona Village Resort. The profusion of petroglyphs depicting sails has led one researcher to suggest that it might indicate a school for sailing or navigation was located here. Although the site is not open to the public, prior arrangements can be made through the social department to view the petroglyphs on the tour offered to guests twice a week.

KALOKO-HONOKŌHAU -- At the Kaloko Honokōhau National Historical Park some younger petroglyphs including a full-rigged ship can be seen. The park can be reached by turning off Highway 19 next to the Honokōhau Harbor exit. A visitor center is located in the park providing maps and interpretive displays. A short trail to Honokōhau Beach from the traihead at the Honokōhau Small Boat Harbor passes a field of petroglyphs.

KAHALU'U -- Some of the most accessible petroglyphs on Hawai'i can be found at Kahalu'u (about five miles south of Kailua-Kona), providing the tide is

Petroglyphs at Kahalu'u, Kona, Hawai'i

out. Seaward of a curving, gray beach about two hundred yards south of the park pavillions is a gently sloping lava flat which is often under water. There is a remarkable variety to the human figures carved there along with phallic symbols and abstract designs. If the pictures are flooded, look along the shore line toward the park and you may find an isolated design.

The Keauhou Archaeological Complex, on the grounds of the adjacent Keauhou Beach Hotel, contains the Ke'ekū Heiau, where Maui's Chief Kamalālāwalu was sacrificed after his defeat. A petroglyph purported to depict him is visible at low tide at the southwest end of the complex.

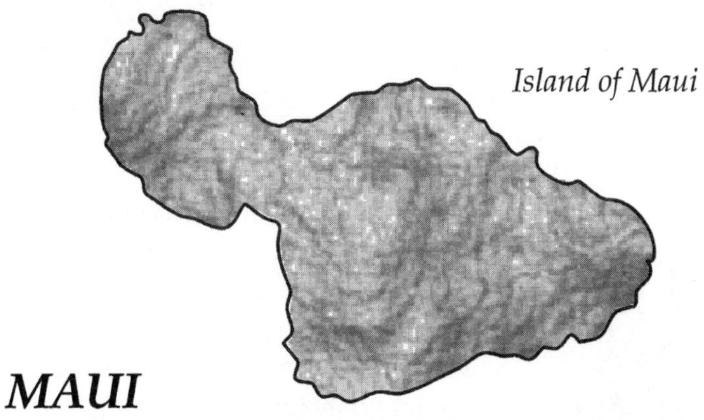

Island of Maui

MAUI

The numerous wide *pahoehoe* flats on the island of Hawai'i were an almost perfect canvas for the people of old. On the other islands in the group, the native artists had to deal with rocks of a different character. They were

obliged to leave their work in shelters, on boulders, and on sandstone flats.

On the sheer bluffs at Nu'u, Hana, Maui, some of the petroglyphs carved into the rock have apparently been painted over with red ocher. In a few places on the island, the painting of symbols was said to be more common. Petroglyphs are found on Haleakala, on West Maui at Lahaina and Kahoma (the thin one), and on Southeast Maui in a shelter cave along Waiohonu Stream.

The collection of petroglyphs at Olowalu is no longer advertised to the public in order to prevent further deterioration and vandalism. A cane field located behind the general store leads to the petroglyphs carved into a cliff face. They are located on private property.

KAHO'OLAWE

Spared from the disruption of development and tourism by its use by the military as "The Target Island," Kaho'olawe has now been returned to the State of Hawai'i. Recent archaeological studies undertaken in conjunction with ordinance cleanup have yielded over 600 newly documented features including a large petroglyph field. Many of the recorded petroglyphs depict goats brought to the islands by Vancouver and given as gifts to Kamehameha. At this time Kaho'olawe is inaccessible to the public.

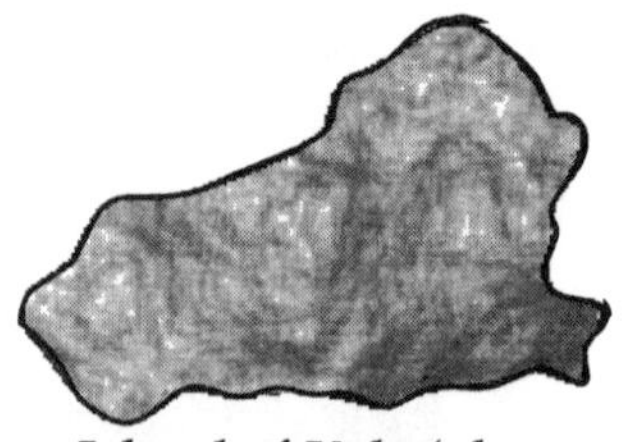

Island of Kaho'olawe

LANA'I

Dense basalt boulders provided the artists' canvas on Lana'i. Petroglyphs are scattered widely over the island, but most of the carvings are represented at Kaunolu and Luahiwa. Site selection is well demonstrated here, where, for example, the boulder Keaohia is covered with drawings while similar boulders around it are bare.

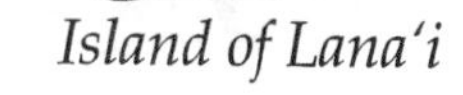

Island of Lana'i

Many images appear to be bird-like, reflecting mythology of the man-eating bird, Halulu, for whom the island's most important shrine, Halulu Heiau is named. The bird-headed figures are especially interesting because here only the head is in profile. These are said to represent humans who anciently had the power to fly. Some families now consider them to be *aumakua* (guardian spirits). Perhaps they might symbolize one of the royalty, who, according to S.M. Kamakau, were entitled to wear a *mahiole* (feathered helmet). Interestingly, the only known pig motif is found on Lana'i.

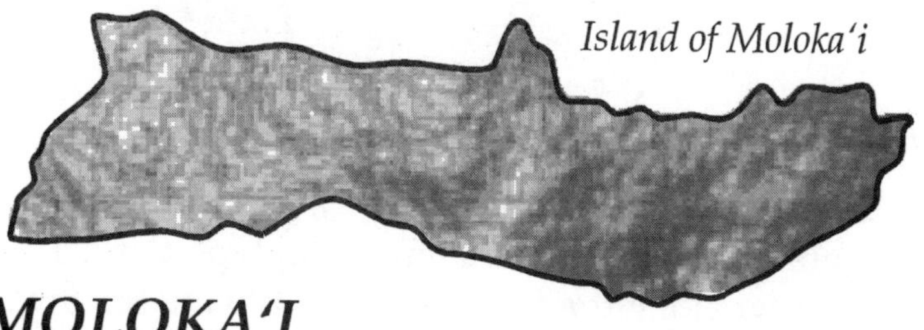

Island of Moloka'i

MOLOKA'I

A provocative story told to John Stokes on the island of Moloka'i concerns a prophetess named Kalaina that lived near Mo'omomi in olden times. One day she went to the trail and carefully scraped and pounded two shallow depressions into the soft sandstone. The next day, she called the people together to see her work. "See

what I have done," she said. "Bye and bye people will come from the sea with feet like these." From that time on the place was called Kalaina *wawae* (Kalaina's feet). According to the narrator, that was the beginning of the fad of visitors leaving their footprints in that place. Of the hundreds of prints there, only one pair has deeper depressions at one end like the heel-marks of boots.

Petroglyphs are also found in a tiny shelter near Ka Ule o Nanahoa, the phallic rock on the seacliff at the northern end of the island.

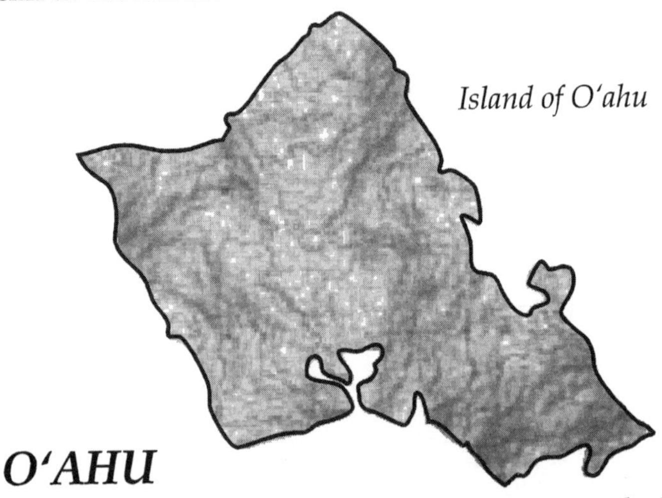

Island of O'ahu

O'AHU

Some of the most accessible Hawaiian rock drawings are protected in Nu'uanu Petroglyph Park. They are located along the west bank of Nu'uanu Stream at Nu'uanu Memorial Park. These carvings may have been inspired by legends of Kaupe, the ghost dog of Nu'uanu.

A large boulder located along the Moanalua Stream in Kamana Nui Valley is known as Kapohakuluahine. It is carved with petroglyphs and a *konane* board. Permission to reach the site is required by the S.M. Damon Estate. Please call to notify them if you plan to hike the trail.

The Koko Head petroglyphs are carved into the floor of a sea cave. Viewing requires climbing on sometimes slippery rocks from a scenic pull-off area on Kalanianaole Highway between Hanauma Bay and Blow Hole. Care also must be taken to be aware of dangerous ocean conditions in this area.

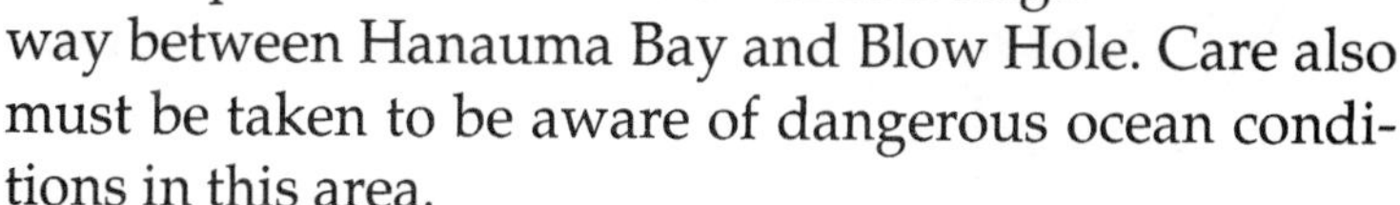

KAUA'I

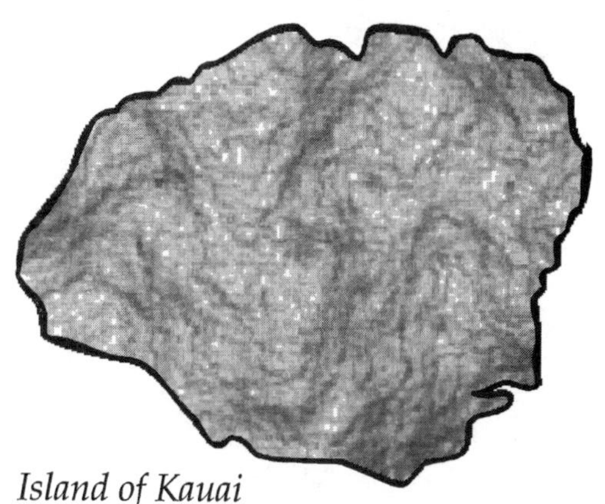

Island of Kauai

Koloa Beach, called in olden times Keoneloa, is located at Mahalepu, east of the sand dunes and Loran station. It is a curious fact that all of the figures on the pictured ledge are said to have their heads toward the sea. Other petroglyphs are found in the tide water below the temples of refuge at the mouth of the Wailua River. At the turn of the century, Mr. Judd reported five carvings at Papalinahoa, Nawiliwili Bay. Two of these lack heads and two others lack arms.

The easiest petroglyphs to see on Kaua'i are the fine representations in the Kaua'i Museum in the city of Lihu'e.

PETROGLYPH ART

The aesthetics of Hawaiian rock drawings have been well summarized by Frances Reed, a long time enthusiast of the local pictographs. "The petroglyph motif, though primitive, is ultra-modern in its appeal. It decorates the interior of the newest jet airplanes that fly among the islands; it appears on fabric used for aloha shirts and *mu'u mu'u*, ceramics and island-printed stationery. It has become another symbol of the romantic aura of a tropical land, along with the palm tree and the hula skirt."

In years past the common practice of making rubbings to record the petroglyph images and simply the tread of many feet over the lava fields have caused damage to this fascinating and mysterious record of the Hawaiians of old. Many petroglyph sites have been defaced by unthinking people using crayons or even paint to emphasize the images. Damage from the use of latex to create molds can also be found. These materials have

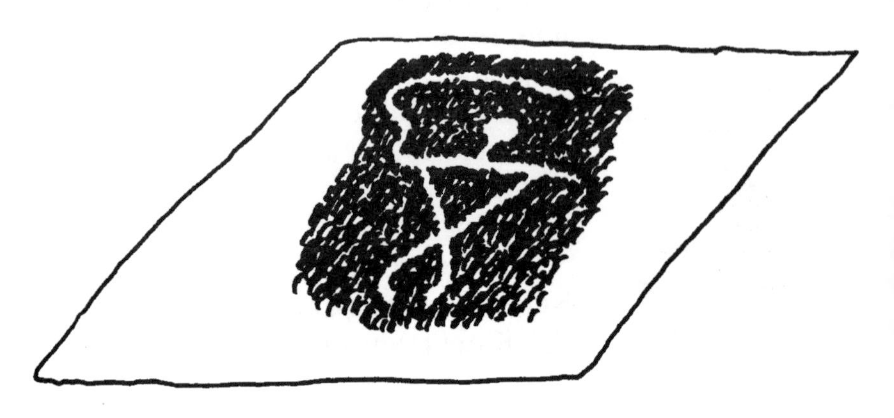

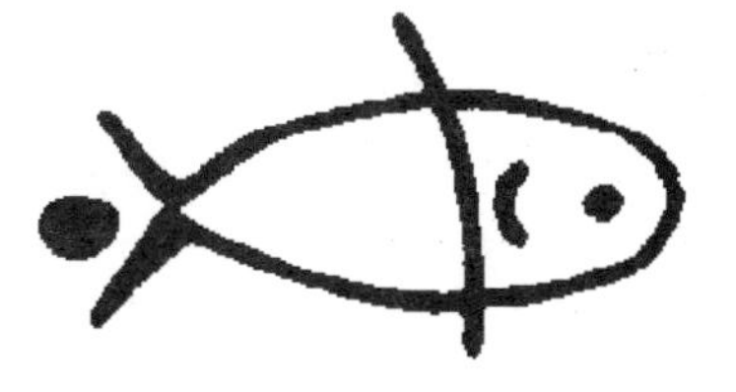

caused irreversible damage to many of the most unique examples of this ancient art. The current historic preservation guidelines for the treatment of petroglyphs is to walk softly and leave them alone!

The Mauna Lani Resort on the Kohala coast of the island of Hawai'i has provided an area at the trailhead to the Puakō Archaeological Preserve with numerous carved reproductions of petroglyph images. This is an ideal place to make rubbings without causing harm to the historic treasures. Please confine any activities other than photography to this area and leave the original petroglyphs undisturbed.

D.K. Reed

These are reproductions of ancient petroglyphs found at the beginning of the trail to the Puakō Petroglyphs. Please confine the making of rubbings to these and preserve the ancient artifacts.

PHOTOGRAPHS -- The high contrast lighting of early morning or late evening is best for capturing the petroglyphs on film. While chalk was often used in the past to obtain contrast, its negative impact can last far into the future.

D.K. Reed

A dense collection of human figures depicted at Puakō.

D.K. Reed

An unusual figure is shown in this Puakō petroglyph.

D.K. Reed

Notice the damage left by a latex mold.

D.K. Reed

This is an example of damage left by an unthinking person.

The following information appears on a sign at the

PUAKŌ PETROGLYPH ARCHAEOLOGICAL DISTRICT

D.K. Reed

Puakō Owl

"Barbed wire and chain link fences are excellent preservation devices, but they detract from the visitor's experience. Through education, and by encouraging enlightened *malama* (stewardship) and pride in the treasures the archaeological district contains, Mauna Lani Resort hopes to enhance and protect the original Hawaiian petroglyph experience for everyone to enjoy.

Since little is known about petroglyphs, their care and handling have come under considerable, often contentious, discussion. If you would like to get involved in the stewardship of the petroglyphs, listed below are resources to help you develop an informed opinion:

Historic Sites Section
Department of Land and Natural Resources
P.O. Box 621, Honolulu, HI 96809
(808) 548-6408

Mauna Lani Resort, Inc.
P.O. Box 4959, Kohala Coast, HI 96743
(808) 885-6677

Rock Art Association of Hawaii (RAAH)
P.O. Box 32902, Honolulu, HI 96837

CONCLUSION

The petroglyphs of Hawai'i are riddles that have intrigued scholars and laymen alike during most of historic time. A few have seen in them the symbol of Siva, the Hindu god; some the beginnings of writing and others perceive only childish native doodles.

As more work is done, the petroglyph puzzle becomes increasingly interesting and more complex. As yet, no single solution has been presented that will answer the many questions the petroglyphs pose. How can we say with certainty that we see in a drawing what a Hawaiian of two hundred years ago would see? Perhaps what appears to us a birth carving was the artist's way of drawing a retainer creeping between the legs of the corpse of a beloved chief to become a companion in death as Kamakau described.

New petroglyph finds and intensive study of those known at present will, in time, add considerably to our knowledge of Hawaiian rock carvings. There is always the hope that a clue will be discovered that will prove to be the key which will unlock the mystery of their age, origin and purpose.

A.S. McBride

Pu'uloa petroglyph which may depict voyaging.

GLOSSARY

ali'i - royalty, those of chiefly blood
ehu - red hair
hala - the pandanus
heiau - temple
kiawe - mesquite
lei - a garland
ka-upu - a gannet or booby bird
kona - the leeward side of an island
kōnane - Hawaiian game similar to checkers
mu'u mu'u - long, flowing dress
pahoehoe - smooth lava
pali - cliff
kapa - bark cloth

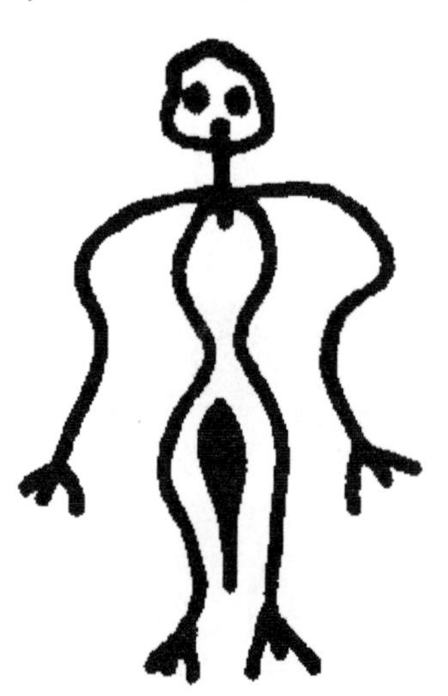

D.K. Reed

Puakō petroglyph, commonly interpreted as a representation of a female, shown as an illustration by the author and as a photo.

BIBLIOGRAPHY

Baker, A.S., *More Petroglyphs, Hawaiian Annual for 1919,* Honolulu

Baker, A.S, *Still More Petroglyphs, Hawaiian Annual for 1920,* Honolulu

Baker, A.S., *Petroglyphs of Ka'u, Hawaiian Annual for 1922,* Honolulu

Baker, A.S., *Puna Petroglyphs, Hawaiian Annual for 1931,* Honolulu

Bennett, W.C, *Archaeology of Kaua'i,* 1931, B.P. Bishop Museum Bulletin 80, Honolulu

Bier, James, *Reference Maps of the Islands of Hawai'i,* 2003, UH Press, Honolulu

Bisignani, J.D., *Hawai'i Handbook,* 1995, Moon Publications, Chico, CA

Buck, Peter H. (Te Rangi Hiroa), *Arts and Crafts of Hawai'i,* 1957, B.P. Bishop Museum, Honolulu

Crowe, Ellie and William, *Exploring Lost Hawai'i,* 2001, Island Heritage, Waipahu, Hawai'i

Cox, J. Halley with Edward Stasack, *Hawaiian Petroglyphs,* 1970, B.P. Bishop Museum, Honolulu

Ellis, William, *A Narrative of a Tour through Hawai'i,* Reprint London ed. 1917

Emory, K.P., *The Island of Lanai,* 1924, B.P. Bishop Museum Bulletin 12, Honolulu

Farley, J.K., *The Pictured Ledge of Kaua'i,* Hawaiian Annual for 1889, Honolulu

Fornander, A., *The Polynesian Race,* 1882, London

Greiner, R.H., *Polynesian Decorative Designs,* 1923, B.P. Bishop Museum Bulletin 7, Honolulu

Holms, R., *New Spain to the Californias by Sea* 1519-1668

Ii, John Papa, *Fragments of Hawaiian History,* 1959, Honolulu

James, Van, *Ancient Sites of Hawai'i,* 1995, Ho'omana'o Arts, Honolulu

James, Van, *Ancient Sites of O'ahu,* 1991, Bishop Museum Press, Honolulu

Kalapana Extension Hawai'i Nat'l. Park, *Natural & Cultural History,* 1959, B.P. Bishop Museum, Honolulu

Kamakau, S.M., *Ka Po'e Kahiko*, 1870, Reprint of Newspaper *Ke Au 'Oko'a*, 1964, Honolulu

Kwiatkowski, P.F., *Na Ki'i Pohaku*, 1991, Ku Pa'a, Inc., Honolulu

Lee, Georgia, *Spirit of Place*, 1999, Easter Island Foundation, Los Osos, CA

Malo, David, *Hawaiian Antiquities*, 1951, B.P. Bishop Museum, Spec. Pub. 2, Honolulu

Mathison, G.F., *Narrative of a Visit to Brazil, Chili, Peru and the Sandwich Islands*, 1822

McAllister, J.G., *Archaeology of O'ahu*, 1933, B.P. Bishop Museum, Bulletin 104, Honolulu

McMahon, Richard, *Adventuring in Hawai'i*, 2003, UH Press, Honolulu

Metraux, A., *Ethnology of Easter Island*, 1940, B.P. Bishop Museum Bulletin 160, Honolulu

Pager, Sean, *Hawai'i Off the Beaten Path*, 1995, Globe Pequot Press, Inc., Old Saybrook, CT

Pukui, Mary Kawena and Samuel H. Elbert, *Hawaiian-English Dictionary*, 1957, Honolulu

Pukui, Mary Kawena and Samuel H. Elbert and Esther T. Mookini, *Place Names of Hawaii*, 1974, UH Press, Honolulu

Reed, Frances, *The Petroglyph Puzzle*, unpublished manuscript

Reeve, Roland B., *Kaho'olawe: Na Leo o Kanaloa*, 1995, 'Ai Pohaku Press, Honolulu

Riegert, Ray, *Hidden Maui*, 2002, Ulysses Press, Berkeley, CA

Stokes, J.F.G., *Notes on Hawaiian Petroglyphs*, 1909, B.P. Bishop Museum, Occasional papers V.4, no. 4, Honolulu

Thrum, Thomas G., *Hawaiian Annual*, 1915

Westervelt, W.D, *Hawaiian Legends of Volcanoes*, 1963, C.E. Tuttle, Co. Reprint, Tokyo

About the Author
Likeke R. McBride

L.R. McBride ~ 1986

The author and illustrator of *Petroglyphs of Hawai'i* lived in Volcano, Hawai'i for over 30 years until his death in October 1993. A student of Hawaiian tradition and culture, McBride also lectured on geology, botany, history and legends of Hawai'i. His death has left a great void.

Of Irish and Iroquois Indian descent, McBride was born in Reading, Pennsylvania and grew up in Ohio. In 1943 he enlisted in the Navy and after basic training, was assigned to a ship stationed at Nawiliwili Harbor, Kaua'i. It was there that he fell in love with the Hawaiian Islands and people and started learning the Hawaiian language and about Hawaiian culture.

After service in World War II and the Korean conflict, McBride received a B.S. degree in geology from Ohio State University, with a minor in botany. After work in industrial research, he joined the National Park Service and was assigned to Hawai'i Volcanoes National Park. With his family, he made his home in the Volcano area. In an eleven year association with Hawai'i Volcanoes National Park, McBride continued to add to his knowledge of all things Hawaiian.

Throughout much of his life McBride was a dedicated student of Hawaiian tradition and culture. He told Hawaiian stories in the old Hawaiian way for nearly thirty years and held a *kauila* dagger, the sign of a professional Hawaiian storyteller. A talented stoneworker and woodcarver, McBride crafted numerous museum-quality reproductions of Hawaiian tools and weapons.

The author wrote and illustrated four other outstanding original works: *About Hawai'i's Volcanoes; The Kahuna, Versatile Masters of Old Hawai'i; Pele, Volcano Goddess of Hawai'i* (currently out of print); and *Practical Folk Medicine of Hawai'i.*

About the Illustrator of the Cover Art
Edwin Kayton

Edwin Kayton is an artist accomplished in many mediums, including oils, life drawing, print-making, etching, sculpture and woodworking. Since moving to Hawai'i in 1977, the primary focus of his artistic endeavors has been the Hawaiian personality with many originals included in corporate collections where they make a profund cultural statement, including Hilton Waikoloa Resort, Outrigger Waikoloa, Hotel King Kamehameha, Mauna Lani Bay Resort, Maui Inter-Continental (Outrigger), Maui Marriott, Kā'anapali Beach Hotel, Outrigger Prince Kūhiō, Hyatt Hotels, Kamehameha Schools and many others. Private collectors include residents in Europe and Japan as well as the U.S.

Kayton studied marble sculpture in Pietrasanta, Italy, in 1994 and continues to live in Italy each summer where he works *plein-air* in the extraordinary atmosphere of this ancient culture.

His originals and reproductions are available at Upcountry Connection Gallery (Waimea) and Lavender Moon Gallery (Kainaliu) as well as on-line: www.kayton-art.com.

The "Petroglyph Maker" by Edwin Kayton.

Notes

Books published by the PETROGLYPH PRESS

ABOUT HAWAII'S VOLCANOES
by L. R. McBride
A CONCISE HISTORY OF THE HAWAIIAN ISLANDS
by Phil K. Barnes
HILO LEGENDS
by Frances Reed
HINA - THE GODDESS
by Dietrich Varez
HOW TO USE HAWAIIAN FRUIT
by Agnes Alexander
JOYS OF HAWAIIAN COOKING
by Martin & Judy Beeman
THE KAHUNA
by L. R. McBride
KONA LEGENDS
by Eliza D. Maguire
LEAVES FROM A GRASS HOUSE
by Don Blanding
PARADISE LOOT
by Don Blanding
PETROGLYPHS OF HAWAII
by L. R. McBride
PLANTS OF HAWAII
by Fortunato Teho
PRACTICAL FOLK MEDICINE OF HAWAII
by L. R. McBride
STARS OVER HAWAII
by E. H. Bryan, Jr.
THE STORY OF LAUHALA
by Edna W. Stall
TROPICAL ORGANIC GARDENING
by Richard L. Stevens

HAWAIIAN ANTIQUITY POSTCARDS
JOHN WEBBER PRINTS

PETROGLYPHS OF HAWAI'I